2	2
3	6
4	24
5	120
6	720
7	5040
8	40320
9	362880
10	3628800
11	39916800
12	479001600
13	6227020800
14	8778291200
15	130767436800
16	20922789888000
17	355687428096000
18	6402373705728000
19	121645100408832000
20	2432902008176640000
21	51090942171709440000
22	1124000727777607680000
23	25852016738884976640000
24	620448401733239439360000

Mutationes, & sic in infinitum .

ABOVE: A 17th-century list of factorial numbers.

First published in the UK in 2021
by Wooden Books Ltd, Glastonbury, UK

Library of Congress Cataloging-in-Publication Data
Linton, O.
Numbers

Library of Congress Cataloging-in-Publication
Data has been applied for

ISBN-10: 1-952178-22-3
ISBN-13: 978-1-952178-22-1

Designed and typeset in Glastonbury, UK

Printed in China on 100% FSC
approved sustainable papers by FSC
RR Donnelley Asia Printing Solutions Ltd.

WOODEN
BOOKS

NUMBERS

TO INFINITY AND BEYOND

Oliver Linton

In memory of my brother Matthew.

Other mathematical titles in the Wooden Books series are: Fractals, On the Edge of Chaos, *by Oliver Linton,* Useful Mathematical & Physical Formulae, *by Matthew Watkins,* Q.E.D., Beauty in Mathematical Proof, *by Burkard Polster,* Ruler and Compass, *by Andrew Sutton, and* The Diagram, *by Adam Tetlow.*

$$1 \times 8 + 1 = 9$$
$$12 \times 8 + 1 = 98$$
$$123 \times 8 + 1 = 987$$
$$1234 \times 8 + 1 = 9876$$
$$12345 \times 8 + 1 = 98765$$
$$123456 \times 8 + 1 = 987654$$
$$1234567 \times 8 + 1 = 9876543$$
$$12345678 \times 8 + 1 = 98765432$$
$$123456789 \times 8 + 1 = 987654321$$

$$9 \times 9 + 7 = 88$$
$$98 \times 9 + 6 = 888$$
$$987 \times 9 + 5 = 8888$$
$$9876 \times 9 + 4 = 88888$$
$$98765 \times 9 + 3 = 888888$$
$$987654 \times 9 + 2 = 8888888$$
$$9876543 \times 9 + 1 = 88888888$$
$$98765432 \times 9 + 0 = 888888888$$
$$987654321 \times 9 - 1 = 8888888888$$

$$1 \times 9 + 2 = 11$$
$$12 \times 9 + 3 = 111$$
$$123 \times 9 + 4 = 1111$$
$$1234 \times 9 + 5 = 11111$$
$$12345 \times 9 + 6 = 111111$$
$$123456 \times 9 + 7 = 1111111$$
$$1234567 \times 9 + 8 = 11111111$$
$$12345678 \times 9 + 9 = 111111111$$
$$123456789 \times 9 + 10 = 1111111111$$

$$1 \times 1 = 1$$
$$11 \times 11 = 121$$
$$111 \times 111 = 12321$$
$$1111 \times 1111 = 1234321$$
$$11111 \times 11111 = 123454321$$
$$111111 \times 111111 = 12345654321$$
$$1111111 \times 1111111 = 1234567654321$$
$$11111111 \times 11111111 = 123456787654321$$
$$111111111 \times 111111111 = 12345678987654321$$

ABOVE: *Decimal sums from sequential inputs of sucessive integers. The beauty of these number pyramids results from using the decimal number base. Choosing specific examples also helps; for instance, replacing the ones with twos in the final palindromic pyramid soon breaks the symmetry!*

CONTENTS

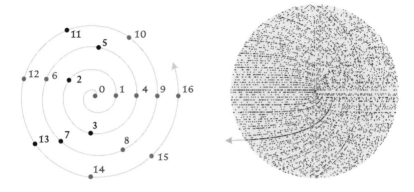

ABOVE: Arrange integers on a spiral, spaced so that an equal length of line is between each. Colour prime numbers in black. As we plot more numbers, patterns appear in the primes, e.g. numbers on the marked curve are of the form $x^2 + x + 41$, the prime-generating formula discovered by Euler in 1772. Images by Robert Sachs.

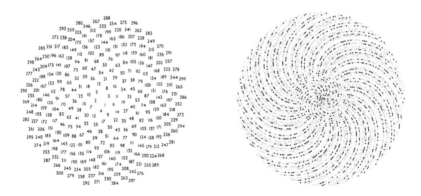

ABOVE: Arrange the numbers on a golden phylotaxis spiral (see page 27). Colour the prime numbers in black, and subprimes in dark grey, etc. As we plot more numbers, some spiral arms appear to host more primes and subprimes than others. Images by Edmund Harrison.

INTRODUCTION

There is a story about the famous Cambridge mathematician G. H. Hardy and his brilliant Indian protégé Srinivasa Ramanujan. Impressed by his work, particularly his uncanny ability to see patterns and relationships in diverse numbers, Hardy had invited Ramanujan over to England to continue his research, but the English weather and the strange food did not agree with him and he was often ill. One day, Hardy took a taxi to the hospital to see his friend. From his bed, Ramanujan asked Hardy the number of the taxi he had come in.

"Oh, it was a very boring number," replied Hardy. "I think it was 1729."

"Actually, that is a very interesting number," said Ramanujan. "It is the smallest number which can be expressed as the sum of two cubes in two different ways." Ever since, numbers which can be expressed as the sums of cubes have been called 'taxicab numbers' (you can find out more about them later in this book).

This story raises an interesting question: are there any *uninteresting* numbers? Well, if there are, there must be one which is 'the smallest uninteresting number', which paradoxically makes it interesting!

Numbers and the patterns they make have fascinated and stimulated mathematical minds for millennia. Sometimes the patterns are fairly trivial, such as the repeated patterns in the illustrations on page iv. Sometimes the patterns hide deep truths, such as the cyclic numbers and Pascal's numbers. Sometimes the patterns are so elusive (like the patterns in the prime numbers) that they have had mathematicians scratching their heads for thousands of years.

Hopefully the patterns revealed in this little book will get you scratching your head—for a little while, at any rate!

WRITING NUMBERS
and the choice of base

We are so familiar with our place-specific decimal notation that it is easy to forget that the collection of symbols 127 is not a number, but a *representation* of a number. The number which it represents is equal to the Egyptian numeral ୨∩∩||||, or the Roman CXXVII, $(1 \times C) + (2 \times X) + (1 \times V) + (2 \times I)$, where C, X, V and I stand for 100, 10, 5, and 1, respectively.

Instead of using different symbols for 10, 100, etc., the Sumerians used a place-specific notation in which the value of a symbol depends on its place in the number. The Sumerian number system was based on 60, so 127 = ∏ ₸ (i.e. $2 \times 60 + 7$), with a small space between the two digits. Paradoxically, they had great difficulty with the number 60 itself because this would be written I (with a small space after it) and was virtually indistinguishable from the number 1. Today we still write numbers so that the first number has the highest value (the so-called *big-endian* system).

Computers represent numbers in *binary*. In this system, only two symbols are used, 0 and 1, represented by low and high voltages (*see below*). Since large binary numbers are rather long, many computer engineers use *hexadecimal* notation, or *hex*, which has a base of 16 (*opposite*). Because 16 is a power of 2 it is easy to turn a hexadecimal number into a binary one and vice versa.

2^7	2^6	2^5	2^4	2^3	2^2	2^1	2^0
128	64	32	16	8	4	2	1
1	0	1	1	0	0	0	1

THE BASE-2 SYSTEM

The binary number
10110001
$= 128 + 32 + 16 + 1$
$= 177$ *decimal*

1	〒	9	〓	30	《《《
2	〓	10	〈	40	〓
3	〓	11	〈〒	50	〓
4	〓	12	〈〓	60	〒
5	〓	13	〈〓	75	〒《〒
6	〓	14	〈〓	100	〒《《
7	〓	15	〈〓	120	〓
8	〓	20	《	961	《〓〒

$$\text{〓 《〈〓 〓 〓}$$
$$(2 \times 60^2) + (23 \times 60) + 40 + 4 = 8624$$

THE SUMERIAN SYSTEM

1	staff
10	cattle hobble
100	coil of rope
1000	lotus flower
10,000	bent finger
100,000	tadpole / frog
1,000,000	Heh, god of the infinite

$$= 12,238$$

EGYPTIAN HIEROGLYPHIC NUMERALS

1	I	9	IX	17	XVII
2	II	10	X	18	XVIII
3	III	11	XI	19	XIX
4	IV	12	XII	20	XX
5	V	13	XIII	50	L
6	VI	14	XIV	100	C
7	VII	15	XV	500	D
8	VIII	16	XVI	1000	M

MMDCLIX = 2,659

THE ROMAN NUMERAL SYSTEM

0	0	8	8
1	1	9	9
2	2	10	A
3	3	11	B
4	4	12	C
5	5	13	D
6	6	14	E
7	7	15	F

$$1D7 = (1 \times 256) + (13 \times 16) + (7 \times 1) = 471$$
$$\dots 65536 \mid 4096 \mid 256 \mid 16 \mid 1$$
$$1 \quad D \quad 7$$

BASE-16 HEXADECIMAL SYSTEM

NATURAL NUMBERS
counting pebbles

The natural numbers 1, 2, 3, 4, ... are the numbers we use from an early age to count a collection of objects, pointing to them one at a time and reciting "one, two, three, four...", etc.

The natural numbers may be *added* and *multiplied* (added and added again a certain number of times) and smaller natural numbers may be *subtracted* from larger ones to produce another natural number.

The Babylonians and the Egyptians had sophisticated ways of adding and multiplying these numbers and subtracting a smaller number from a larger one, but neither civilisation had any concept of negative numbers (debts were accounted separately from assets), nor of zero.

To the ancient Greeks, numbers like 10 or 21 with relatively few divisors were called *deficient* because the sum of their divisors is less than the number itself, e.g. the factors of 10 are 1, 2, and 5, which sum to 8. Other numbers, like 12 and 20, have many factors and were called *abundant*, because the sum of their divisors is greater than the number itself, e.g. the factors of 12 are 1, 2, 3, 4, and 6, which sum to 16. Between these two extremes, a number was considered *perfect* if it was equal to the sum of its divisors, e.g. the factors of 6 are 1, 2, and 3, which sum to 6, or those of 28 are 1, 2, 4, 7, and 14, which sum to 28.

The simplest and most obvious pattern in the sequence of natural numbers is the division into alternate odd and even numbers (*below*).

EVEN × EVEN = EVEN	EVEN ± EVEN = EVEN
EVEN × ODD = EVEN	EVEN ± ODD = ODD
ODD × ODD = ODD	ODD ± ODD = EVEN

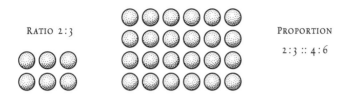

RATIO 2:3

PROPORTION

2:3 :: 4:6

ABOVE: RATIO and PROPORTION. A ratio occurs between two numbers. A proportion then occurs between two ratios. These two rectangles have the same proportions because 2 is to 3 as 4 is to 6.

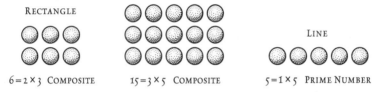

RECTANGLE

LINE

6 = 2 × 3 COMPOSITE

15 = 3 × 5 COMPOSITE

5 = 1 × 5 PRIME NUMBER

ABOVE: The Pythagoreans were fascinated by RECTANGULAR NUMBERS. Any number which can be expressed as a rectangle (as opposed to a line) can be divided by the side and the width, so is a COMPOSITE NUMBER. Any number which cannot be expressed as a rectangle is a PRIME NUMBER.

		5 times 1	5
		times 2	10
		times 3	15
		times 4	20
		times 5	25
		times 6	30
		times 7	35
		times 8	40
		times 9	45
		times 10	50
		times 11	55

LEFT: A Babylonian 5 × multiplication table, c. 1750 BC. From tablet NBC7344 in the Yale Babylonian Collection.

Methods of multiplication were known in ancient Egypt, India, Greece, and China. The Egyptian method used successive additions and doublings, e.g.

$$13 \times 21 = (1 + 4 + 8) \times 21$$
$$= (1 \times 21) + (4 \times 21) + (8 \times 21)$$
$$= 21 + 84 + 168 = 273.$$

Composite numbers are formed by the multiplication of smaller integers.

PRIME NUMBERS
and their distribution

Any number that cannot be formed by multiplying two smaller numbers, each larger than 1, is said to be a *prime number*. For example, 7 is prime because the only way of writing it as a product is 1×7.

Two numbers are said to be *coprime* if the only common whole number that divides them is 1, e.g. the numbers 4 and 9 are coprime—if we draw a 4×9 lattice, the diagonal will not intersect any other points (*below*).

The Greek mathematician Euclid [c.350–250 BC] proved that there are an infinite number of primes. Suppose there is a largest prime, he said. Now consider the huge number made by multiplying all the primes together and adding 1. This cannot be divided by any of the primes because there will always be a remainder 1, so either it must be prime or it must have at least one prime factor larger than any in our list. So the list cannot be complete—the number of primes must be infinite.

While it is obvious that the distribution of primes is not random, it is very difficult, if not impossible, to find any regular pattern in the way they are distributed. However, strange patterns do exist (*see Ulam's Spiral, opposite*). In 1792 the 15-year-old Carl Friedrich Gauss conjectured that the density of primes decreased according to a logarithmic law (*opposite*) but although his conjecture is remarkably accurate, the precise way in which the primes are distributed remains unknown.

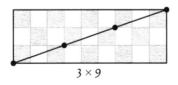

3×9

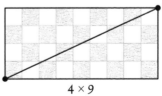

4×9

1	2	3	4	5	6	7	8	9	10
11	12	13	14	15	16	17	18	19	20
21	22	23	24	25	26	27	28	29	30
31	32	33	34	35	36	37	38	39	40
41	42	43	44	45	46	47	48	49	50
51	52	53	54	55	56	57	58	59	60
61	62	63	64	65	66	67	68	69	70
71	72	73	74	75	76	77	78	79	80
81	82	83	84	85	86	87	88	89	90
91	92	93	94	95	96	97	98	99	100

ABOVE: The numbers from 1–100, with the primes highlighted. All the other numbers are composite numbers, and can be formed as the product of two or more primes.

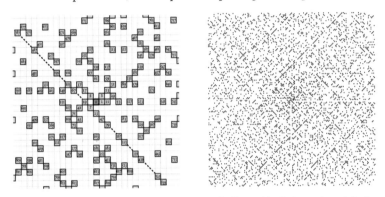

ABOVE: ULAM'S SPIRAL. In 1963, Stanislaw Ulam was doodling, writing the integers in a spiral. Idly, he highlighted all the prime numbers and was astonished to find diagonal lines cropping up all over the place. The reason for these diagonals remains a mystery. A similar pattern is shown facing page 1.

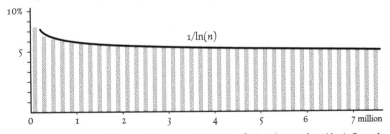

ABOVE: THE PRIME NUMBER THEOREM. The density of primes decreases logarithmically. In the region of $n = 1$ million, there are approx. 7 primes per 100 numbers. $(100 \times 1/\ln(1000000) \approx 7)$

POLYGONAL NUMBERS

triangles and squares

Arranging pebbles into triangles, squares, and other shapes can reveal some fascinating relationships. For example, the nth triangular number is the sum of all the natural numbers up to n.

There is a story that the eight-year old Gauss [1777–1855] and his class were asked by their teacher to add up all the numbers from 1 to 100 (which would give the 100th triangular number). Almost before the class had settled down to work, Gauss brought her a piece of paper with just a single number on it—5050. Astonished, she asked him how he did it. Gauss explained:

$$1 + 2 + 3 + \cdots + 50 + 51 + \cdots + 98 + 99 + 100$$
$$= (1 + 100) + (2 + 99) + (3 + 98) + \cdots + (50 + 51)$$
$$= 101 + 101 + 101 + \cdots + 101$$
$$= 50 \times 101 = 5050.$$

So, Gauss's formula for the nth triangular number $= \frac{n}{2} \times (n + 1)$.

The nth *square number* (n^2), meanwhile, is equal to the sum of the first n odd numbers, and also equal to the sum of the $(n-1)$th and the nth triangular numbers (*see opposite*).

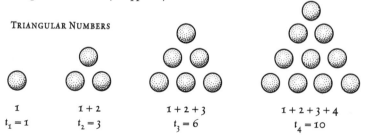

TRIANGULAR NUMBERS

1	1 + 2	1 + 2 + 3	1 + 2 + 3 + 4
$t_1 = 1$	$t_2 = 3$	$t_3 = 6$	$t_4 = 10$

SQUARE NUMBERS

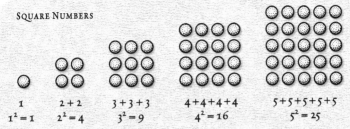

$$1 \quad\quad 2+2 \quad\quad 3+3+3 \quad\quad 4+4+4+4 \quad\quad 5+5+5+5+5$$
$$1^2 = 1 \quad\quad 2^2 = 4 \quad\quad 3^2 = 9 \quad\quad 4^2 = 16 \quad\quad 5^2 = 25$$

ABOVE: *The first five square numbers. These were intensively studied by the ancient Pythagoreans, as were the triangular numbers, shown on the facing page, with 10 being particularly revered.*

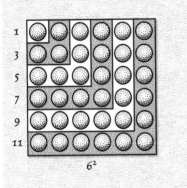

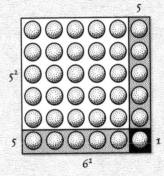

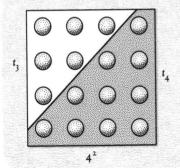

PATTERNS IN SQUARE NUMBERS

ABOVE LEFT: *Every square number n^2 is the sum of the first n odd numbers. Example: $4^2 = 16 = 1 + 3 + 5 + 7$.*

ABOVE RIGHT: *By the same reckoning, if we know n^2, then $(n+1)^2 = n^2 + 2n + 1$.*

LEFT: *Every square number n^2 is the sum of the $(n-1)$th and the nth triangular numbers. Example: $4^2 = 16 = 6 + 10$.*

POLYHEDRAL NUMBERS
apples and oranges

Arranging oranges or cannon balls into cubes, pyramids, and other shapes reveals some fascinating relationships between triangular numbers, square numbers and cubic numbers.

For example, although it may be easy to see why the nth tetrahedral number will be the sum of the first n triangular numbers, or the nth pyramidal number will be the sum of the first n square numbers (*below*), it may be more surprising to discover that the sum of all the cubic numbers up to n is equal to the square of the nth triangular number (*see visual proof, opposite*).

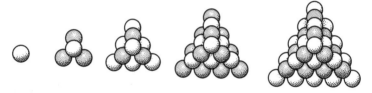

TETRAHEDRAL NUMBERS: $1, 4, 10, 20, 35 ... N = n(n+1)(n+2)/6$
These are the sum of the first n triangular numbers.

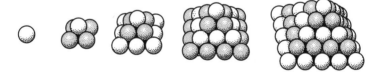

PYRAMID NUMBERS: $1, 5, 14, 30, 55 ... N = n(n+1)(2n+1)/6$
These are the sum of the first n square numbers.

$$1^3 = 1 = 1$$
$$2^3 = 8 = 3 + 5$$
$$3^3 = 27 = 7 + 9 + 11$$
$$4^3 = 64 = 13 + 15 + 17 + 19$$
$$5^3 = 125 = 21 + 23 + 25 + 27 + 29$$
$$6^3 = 216 = 31 + 33 + 35 + 37 + 39 + 41$$
$$7^3 = 343 = 43 + 45 + 47 + 49 + 51 + 53 + 55$$
$$8^3 = 512 = 57 + 59 + 61 + 63 + 65 + 67 + 69 + 71$$
$$9^3 = 729 = 73 + 75 + 77 + 79 + 81 + 83 + 85 + 87 + 89$$

LEFT: CUBIC NUMBERS. *Every cubic number is the sum of a consecutive sequence of odd numbers. This simple observation has an even more remarkable consequence, known as Aryabhata's theorem after the Indian mathematician and astronomer who discovered it in 499 AD.*

$$1^3 + 2^3 = 9 = 3^2 = (1 + 2)^2$$
$$1^3 + 2^3 + 3^3 = 36 = 6^2 = (1 + 2 + 3)^2$$
$$1^3 + 2^3 + 3^3 + 4^3 = 100 = 10^2 = (1 + 2 + 3 + 4)^2$$
$$1^3 + 2^3 + 3^3 + 4^3 + 5^3 = 225 = 15^2 = (1 + 2 + 3 + 4 + 5)^2$$
$$1^3 + 2^3 + 3^3 + 4^3 + 5^3 + 6^3 = 441 = 21^2 = (1 + 2 + 3 + 4 + 5 + 6)^2$$

LEFT: ARYABHATA'S THEOREM. *The sum of the cubes up to n = the square of the nth triangular number, i.e. the square of the sum of the first n numbers.*

CUBIC NUMBERS:
$$1, 8, 27, 64, 125, \ldots N = n^3$$

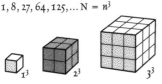

1^3 2^3 3^3 4^3 5^3

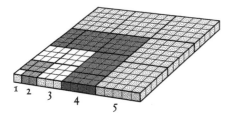

1   2   3   4   5

LEFT: Visual proof of Aryabhata's Theorem. Each cube is cut into slices, and the top layer of even-numbered cubes is then cut in half. The slices are then all assembled to form the complete square.

PYTHAGOREAN TRIPLES
and Fermat's last theorem

Egyptian architects knew that rods of length 3, 4, and 5 cubits would form a right-angled triangle, and finding other integer triples which obeyed the rule $a^2 + b^2 = c^2$ was set as an exercise to Babylonian schoolchildren. Such triplets of integers are known as *Pythagorean triples*. Euclid discovered that you could generate them all using the formula below, where m and n are integers with $m > n$:

$$a = m^2 - n^2 \quad b = 2mn \quad c = m^2 + n^2$$

The illustrations opposite show plots of Pythagorean triples with numbers under 30 (*upper*) and a scatter plot of those with numbers under 500 (*lower*), showing how the dots lie on two sets of related parabolas.

What happens when we try to use cubes instead of squares in Pythagoras' Theorem? Pierre de Fermat [1607–1665] noted in the margin of a book that he had found a "truly remarkable proof" that no solutions existed to the equation $a^n + b^n = c^n$ where $n > 2$ and a, b, and c are positive integers. In 1995, Andrew Wiles and others proved the theorem correct.

Although no cube numbers can be the sum of two other cubes, they can be the sum of three. For example: $2^3 + 17^3 + 40^3 = 41^3$. Three fourth powers can also add up to a fourth power. The smallest solution was found in 1987: $95800^4 + 217519^4 + 18796760^4 = 20615673^4$.

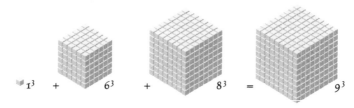

$1^3 \quad + \quad 6^3 \quad + \quad 8^3 \quad = \quad 9^3$

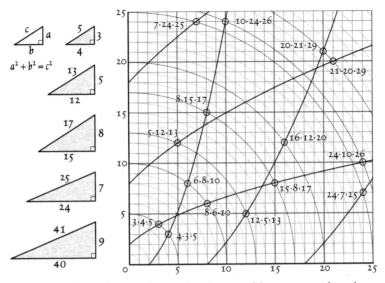

$$a^2 + b^2 = c^2$$

Triangles shown: $5, 4, 3$; $13, 12, 5$; $17, 15, 8$; $25, 24, 7$; $41, 40, 9$

Graph labels: $7\cdot24\cdot25$, $10\cdot24\cdot26$, $20\cdot21\cdot29$, $21\cdot20\cdot29$, $8\cdot15\cdot17$, $5\cdot12\cdot13$, $16\cdot12\cdot20$, $24\cdot10\cdot26$, $6\cdot8\cdot10$, $15\cdot8\cdot17$, $24\cdot7\cdot25$, $3\cdot4\cdot5$, $8\cdot6\cdot10$, $12\cdot5\cdot13$, $4\cdot3\cdot5$

ABOVE LEFT: Simple Pythagorean triangles, such as the 3-4-5 and the 5-12-13, are early members of a fascinating series. ABOVE RIGHT: Each Pythagorean triple occurs where the integer circles exactly cross an intersection in the grid. The parabolas connect triples with the same value of n. For example, n=1 and m=2 gives 3,4,5, and n=1 and m=3 gives 8,6,10.

RIGHT: A zoomed-out version of the same diagram with a and b less than 500. Each Pythagorean triplet is shown as a small dot.

FACING PAGE: An example of one cube being the sum of three other cubes, $9^3 = 1^3 + 6^3 + 8^3$. On page 1 we met Ramanujan's 'taxicab' number 1729, the smallest number which can be expressed as the sum of two cubes in two different ways. We may also find the smallest number which is the sum of two squares in two different ways. e.g. $65 = 7^2 + 4^2 = 8^2 + 1^2$, and $125 = 11^2 + 2^2 = 10^2 + 5^2$.

ZERO
much ado about nothing

Is zero a number? If so, is it positive or negative? Is it even or odd? Can you multiply by zero? Can you divide by zero? Can you raise a number to the power of zero? Does this operation actually make any sense? The Greek mathematician Diophantus [c.210–c.290 AD] thought that the whole idea of zero and negative numbers was absurd.

One of the most important things to grasp about mathematics is that (with the possible exception of the natural numbers) you can define any symbol or operator that you use to mean anything that you like. If I want, I can define 2×0 to be 33, or 5 divided by zero to be -1, or 7 to the power zero to be green cheese. However, if I want to construct a mathematics in which the rules can be applied universally and consistently, then we are severely constrained as to the definitions we can adopt. For example, we want $2 \times (3 + 0)$ to be the same as 2×3. But applying what is known as the distributive law we get $2 \times (3 + 0) = 2 \times 3 + 2 \times 0$. So 2×0 must be equal to zero. Similar arguments show that the product of two negative numbers must be a positive number.

But when it comes to division it turns out that there is no consistent way in which you can define 5 divided by zero to be anything meaningful. Any number divided by zero is technically undefined and the operation of dividing by zero is banned.

7 to the power zero can be defined in a meaningful way (*see opposite*). You can even define raising to the power of a negative number.

Later we shall come across *factorial* numbers. Factorial 6 (written 6!) $= 1 \times 2 \times 3 \times 4 \times 5 \times 6 = 720$. Factorial $1 = 1$. But what is factorial 0? Surprisingly, it turns out that we have to define it as being 1 also.

THE ZEROTH POWER - IN SUGAR CUBES

A. 5 cubed is $5^3 = 5 \times 5 \times 5 = 125$, where 3 is the exponent.

B. Divide the stack of cubes into 5 slices. Each $5^2 = 5 \times 5 = 25$ slice is equal to $125 \div 5$.

C. Divide the slice again into 5 lines. Each line has 5 cubes. And $5^1 = 5 = 25 \div 5$.

(Note that every time we reduce the exponent by 1, the number of cubes is divided by 5)

D. Carry on and divide the line into 5 cubes. For a single cube the exponent is now zero. So, to be consistent, $5^0 = 1$. Q.E.D.

In a similar way we must conclude that $5^{-1} = 1/5$; $5^{-2} = 1/5^2$; $5^{-3} = 1/5^3$; etc.

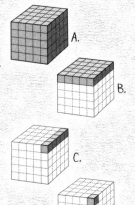

A.

B.

C.

D.

THE FACTORIAL OF ZERO

The factorial of a number, n!, is the product of all the positive integers up to and including that number.

EXAMPLE: $6! = 1 \times 2 \times 3 \times 4 \times 5 \times 6$.

So why do mathematicians tell us that the factorial of zero is one?

We know that $1! = 1$ and we also know that $n! = n \times (n-1)!$

Let's substitute $n = 1$

$1! = 1 \times (1-1)!$ so... $1 = 1 \times 0!$

So 0! must be equal to 1. Q.E.D

$$0! = 1$$

Negative Numbers
putting your ducks in a row

Every mathematical operation has its inverse, and applying inverse operations can lead to new kinds of numbers. For example, the inverse of addition is subtraction, and although the ancient Greeks thought the idea of subtracting 8 from 3 was absurd, many centuries later *negative* quantities became accepted as genuine numbers.

It is difficult to pin down exactly when or where negative numbers first became accepted. By the 12th century, Indian and Arabian mathematicians had worked out the rules for the addition and multiplication of negative numbers (although zero was still held in suspicion). But it was not until René Descartes [1596–1650] published his seminal book *La Géométrie* in 1637 that it became clear that numbers were best represented by points along a line—positive numbers to the right and negative numbers to the left—stretching off to infinity in both directions with zero in the middle (*shown below*). Adding is then simply the operation of moving to the right; subtraction is simply the operation of moving to the left so, for example, 2 − 3 = −1 (*opposite, bottom*).

The natural numbers (*see page 4*) plus zero and the negative numbers are known as the *integers* and form what is known as a *group* under addition (and its inverse, subtraction). This means you can add or subtract any pair of integers and you will always get another integer.

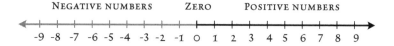

NEGATIVE NUMBERS   ZERO   POSITIVE NUMBERS

-9 -8 -7 -6 -5 -4 -3 -2 -1 0 1 2 3 4 5 6 7 8 9

FUNDAMENTAL PROPERTIES OF ALL NUMBERS

PROPERTY	ADDITION	MULTIPLICATION
COMMUTATIVE	$a + b$ = $b + a$	$a \times b$ = $b \times a$
ASSOCIATIVE	$a + (b + c)$ = $(a + b) + c$	$a \times (b \times c)$ = $(a \times b) \times c$
DISTRIBUTIVE	n/a	$a \times (b + c)$ = $(a \times b) + (a \times c)$
IDENTITY	$a + 0 = a$	$a \times 1 = a$
INVERSE	$a + (-a) = 0$	$a \times \frac{1}{a} = 1$

ABOVE: The integers form a group under addition, subtraction, and multiplication. Including division in the list of operators requires including all the rational fractions too (i.e. all those numbers which can be expressed as a ratio of two integers.) FACING: The NUMBER LINE of negative and positive numbers. BELOW: Addition and subtraction can be thought of as movements along the number line.

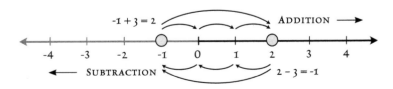

DIVIDED NUMBERS
fractions, remainders and decimals

The opposite of multiplication is division. This is easy if you are dividing a number by one of its factors, but can become more complex otherwise. One way to deal with a tricky division is simply to leave a *remainder*, the basis of *modular arithmetic* (*see page 20*). Another way is to create a *fraction*. But how best to write this? As with zero and negative numbers, it took centuries for mathematicians to realise that there were numbers between the integers, and how best to represent them.

The Babylonians had a set of symbols to represent some fractions ($\frac{1}{2}$, $\frac{2}{3}$, $\frac{5}{6}$, etc.) but lacked a general framework. The Egyptian system was more flexible, but only used unit fractions ($\frac{1}{2}$, $\frac{1}{3}$, $\frac{1}{4}$, $\frac{1}{5}$, etc) and represented numbers like $\frac{2}{3}$ as sums of these, $\frac{1}{2} + \frac{1}{6}$.

Today, we still use two different systems to write non-integer numbers: fractions and *decimals*, our common base-10 place-value system. In base 10, if the denominator of a fraction does not divide into a power of 10 then the division process never ends and we get what is known as a *recurring decimal*. For example, the fraction $\frac{2}{3}$ in decimal comes out as 0.666666... recurring. Opposite we see what happens when we build a decimal expansion for $\frac{1}{7}$. The base you use determines which fractions have finite expansions and which ones are recurring. For example, 0.1 in decimal becomes an infinitely recurring expansion in binary.

$$\frac{1}{2} \quad + \quad \frac{1}{3} \quad = \quad \frac{3}{6} \quad + \quad \frac{2}{6} \quad = \quad \frac{5}{6}$$

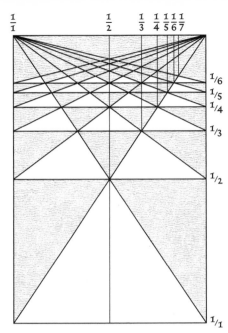

$$\frac{1}{1} \qquad \frac{1}{2} \qquad \frac{1}{3} \ \frac{1}{4}\frac{1}{5}\frac{1}{6}\frac{1}{7}$$

$1/6$

$1/5$
$1/4$

$1/3$

$1/2$

$1/1$

LEFT: The famous Diagram of Villard de Honnecourt [1200–1250] which shows a simple and precise geometric construction for any rational fraction.

FACING: Fractions are added and subtracted using a common denominator.

Fractions are multiplied as:

$$\frac{a}{b} \times \frac{c}{d} = \frac{ac}{bd}$$

and divided by inverting the divisor and then multiplying:

$$\frac{a}{b} \div \frac{c}{d} = \frac{a}{b} \times \frac{d}{c} = \frac{ad}{bc}$$

NOTE: Non-integer numbers can be written in any base, e.g. hex 123.AB in decimal is $1 \times (16)^2 + 2 \times (16)^1 + 3 \times (16)^0 + 10 \times (16)^{-1} + 11 \times (16)^{-2}$
$$= 256 + 32 + 3 + \frac{10}{16} + \frac{11}{256}$$
$$= 291.66796875$$

$1/7 = 0.\ 142857\ 142857... = 0.\dot{1}4285\dot{7}$

$2/7 = 0.\ 285714\ 285714... = 0.\dot{2}8571\dot{4}$

$3/7 = 0.\ 428571\ 428571... = 0.\dot{4}2857\dot{1}$

$4/7 = 0.\ 571428\ 571428... = 0.\dot{5}7142\dot{8}$

$5/7 = 0.\ 714285\ 714285... = 0.\dot{2}8571\dot{4}$

$6/7 = 0.\ 857142\ 857142... = 0.\dot{8}5714\dot{2}$

ABOVE: Sevenths shown as decimals. All use repeats of rotations of the same six numbers. The dots indicate the pattern of the repeats.

BELOW: If you divide 1 by 7 longhand in base 10, the only remainders are the numbers 1 to 6:

$$7\ \underline{|\ 1.\ {}^1 0\ {}^3 0\ {}^2 0\ {}^6 0\ {}^4 0\ {}^5 0\ {}^1 0 ...}$$
$$0.\ 1\ 4\ 2\ 8\ 5\ 7\ 1\ ...$$

$1/3 = 0.\ 333333.... = 0.\dot{3}$

$1/6 = 0.\ 166666.... = 0.1\dot{6}$

$1/7 = 0.\dot{1}4285\dot{7}$

$1/9 = 0.\ 111111... = 0.\dot{1}$

$1/11 = 0.0909... = 0.\dot{0}\dot{9}$

$1/13 = 0.\dot{0}7692\dot{3}\dot{0}$

$1/14 = 0.0\dot{7}1428\dot{5}$

$1/15 = 0.066666 = 0.06\dot{6}$

$1/17 = 0.\dot{0}58823529411764\dot{7}$

$1/19 = 0.\dot{0}5263157894736842\dot{1}$

ABOVE: Repeating decimal versions of some simple fractions, e.g. the decimal for $\frac{1}{17}$ repeats every 16 digits.

Modular Numbers
counting in circles

There is a famous problem in number theory, originally derived from calendars (*see opposite*). Suppose there are N stepping stones arranged in a circle labelled from zero to $N-1$. Starting from Stone zero and moving round in steps of S, the problem is to determine whether it is possible to reach Stone R and, if so, how many steps it will take.

The illustrations below show the paths that you will take if $N = 12$ and $S = 1, 2, 3, 4, 5,$ and 6. Note that whenever N is divisible by S, some of the stepping stones are unreachable. It is not quite so obvious but it is easy to prove that if N and S have no common divisors (i.e. they are coprime), then all the stones are accessible. Take, for example, the situation when $N = 27$ and $S = 10$ (*see opposite*).

The general rule is that Stone R will be reachable from stone zero if and only if R is divisible by the greatest common divisor of N and S.

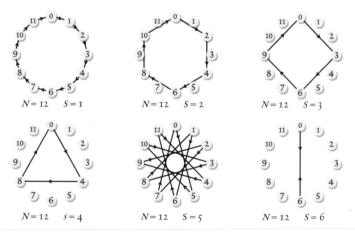

$N = 12 \quad S = 1$ $N = 12 \quad S = 2$ $N = 12 \quad S = 3$

$N = 12 \quad s = 4$ $N = 12 \quad S = 5$ $N = 12 \quad S = 6$

LEFT: MODULAR ARITHMETIC *goes back thousands of years and can assist with particular types of questions involving cycles and synchronisations. For example: "When will Christmas Day next fall on a Thursday?" or "When will the full moon next fall on the first day of the month?" This kind of arithmetic was first formally systematised by Carl Friedrich Gauss in 1801.*

LEFT: *Cyclic problems involve solving the Diophantine equation $aS = bN + K$ (where N is the cycle period, S is the step size and K is a constant) for integers a and b. This is the simplest possible Diophantine equation and was studied by the Alexandrian mathematician Diophantus [c.210-290AD]. The equation is only guaranteed to have a solution if S and N are coprime. For example, Christmas Day can fall on any day of the week because 7 and 365 are coprime.*

DIOPHANTINE EQUATIONS

have the form: $aS = bN + K$

where N is the cycle period,
S is the step size, and
K is a constant for integers $a + b$

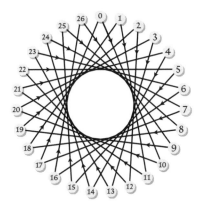

LEFT: *The path generated using 27 stones and a step of 10. We want to find n, the number of steps we need to reach Stone 8. In other words, 10n leaves a remainder of 8 when divided by 27. Or, as a modular equation:*

$$10n \equiv 8 \ (mod \ 27)$$

To find n, just keep adding 27 to the right-hand side until a multiple of 10 is reached:

$$8 \rightarrow 35 \rightarrow 62 \rightarrow 89 \rightarrow 116 \rightarrow 143 \rightarrow 170$$

so, $10n \equiv 170 \ (mod \ 27)$

Dividing both sides by 10 we find that n = 17 (Check: 170 ÷ 27 = 6 remainder 8)

Rational Numbers

how many are there?

How many even numbers are there? And how many square numbers?

It turns out that for every even number there is a corresponding natural number (its half) and for every square number there is a natural number (its square root) so the *cardinality* (quantity) of the even numbers and the square numbers is the same as that of the natural numbers, i.e. *infinity*, written ∞.

How many rational fractions are there? It turns out that for every fraction there is a natural number (*see opposite*), so their cardinality is the same too. Ordered triples (e.g. 1-2-3) also have the same cardinality, as they can be counted using a pyramidal arrangement, and this can be extended to any number of dimensions. Put them all together and you still get infinity, as ∞ + ∞ = ∞ and ∞² = ∞ (*see page 40*).

The *rational fractions* can be expressed as $^a/_b$ (where *a* and *b* are whole numbers). Together with the integers and their negative counterparts these comprise the *rational numbers*, which form a *group* (*see page 17*) under addition, subtraction, multiplication and division. For any pair of rationals we can find at least one rational fraction which lies between them, and since there must be at least one fraction in the spaces between this fraction and the two original ones, it follows that there must be an infinite number of rational fractions between 0 and 1 and therefore between any two integers (*see opposite, and below*).

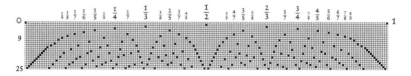

Q: How many	TRI	1	2	3	4	5	6	7
FRACTIONS	0	1/2						
are there?	1	1/3	2/3					
	3	1/4	2/4	3/4				
A: One for	6	1/5	2/5	3/5	4/5			
every natural	10	1/6	2/6	3/6	4/6	5/6		
number!	15	1/7	2/7	3/7	4/7	5/7	6/7	
	21	1/8	2/8	3/8	4/8	5/8	6/8	7/8

ABOVE: A table of fractions, summed using the triangular numbers. Amazingly, every fraction is uniquely associated with a natural number, e.g. $4/7$ is the 19th fraction in the list $(= 15 + 4)$. So the cardinality of the rational fractions is the same as the cardinality of the natural numbers.
BELOW: An infinite number of fractions lie between any two integers.

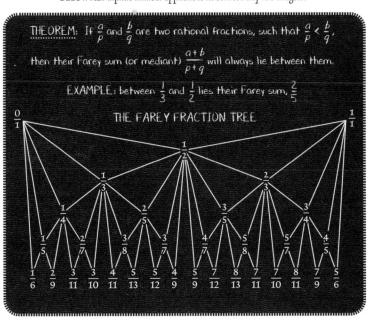

THEOREM: If $\frac{a}{p}$ and $\frac{b}{q}$ are two rational fractions, such that $\frac{a}{p} < \frac{b}{q}$, then their Farey sum (or mediant) $\frac{a+b}{p+q}$ will always lie between them.

EXAMPLE: between $\frac{1}{3}$ and $\frac{1}{2}$ lies their Farey sum, $\frac{2}{5}$

THE FAREY FRACTION TREE

IRRATIONAL NUMBERS
the root of the matter

Irrational numbers cannot be expressed as simple fractions. When Hippasus [530–450 BC] discovered that the length of the diagonal of a 1 × 1 unit square was not a rational fraction, his fellow Pythagoreans are reputed to have drowned him at sea to keep it quiet.

We can easily prove that $\sqrt{2}$ is irrational. Suppose that $\sqrt{2} = p/q$, where p and q are integers without any common factors. Since $p^2/q^2 = 2$, it follows that $p^2 = 2q^2$, so p^2 must be even and p must be even too, i.e. $p = 2n$ for some integer n, so $2q^2 = 4n^2$, or $q^2 = 2n^2$. But this means that q must be even as well. This contradicts our original assumption that p and q have no common factors, so $\sqrt{2}$ cannot be expressed as a rational fraction.

It can easily be shown that if the nth root of an integer is not itself an integer, it cannot be expressed as a rational fraction. In short, it must be irrational and hence the diagonal of any rectangle with integer sides is either integer or irrational. Irrational numbers have a never-ending, non-repeating decimal expansion; indeed, they have an infinite number of digits after the point in any base.

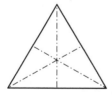

Altitude of a triangle
of side $1 = \frac{1}{2}\sqrt{3}$

Diagonal of a square
of side $1 = \sqrt{2}$

Diagonal of a pentagon
of side $1 = 2/(\sqrt{5} - 1)$

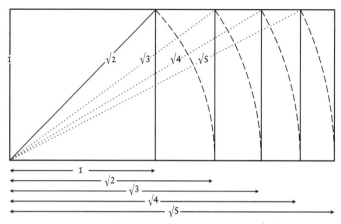

ABOVE: *Irrational rectangles. Starting with a square, we swing the $\sqrt{2}$ diagonal down to produce a $\sqrt{2}$ rectangle, whose diagonal is $\sqrt{3}$. Swinging this down produces a $\sqrt{3}$ rectangle, whose diagonal is $\sqrt{4} = 2$. Swinging this down produces a $\sqrt{2}$ rectangle, whose diagonal is $\sqrt{5}$.*

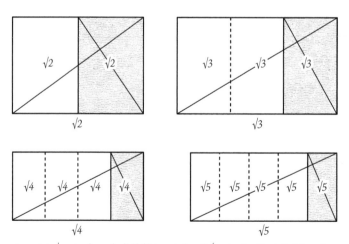

ABOVE: *A $\sqrt{n}$ rectangle may be divided into n equal small $\sqrt{n}$ rectangles. The A-size paper series are all $\sqrt{2}$ rectangles, so when you fold a piece of A4 paper in half, you obtain another $\sqrt{2}$ rectangle, A5.*

THE GOLDEN RATIO
rabbits and cows

In the 12th century, the Pisan mathematician Fibonacci posed this problem: Female rabbits take 2 months to mature and thereafter produce 1 female rabbit every month. How does the population grow?

Starting with 1 female in month 1, we still have 1 female in month 2. In months 3 and 4 she gives birth to a 2nd and 3rd female. By the 5th month there are 2 mature females so the population jumps to 5, then 8, and so on. This is the famous Fibonacci sequence:

$$1, 1, 2, 3, 5, 8, 13, 21, 34, 55, \ldots$$

Note how each successive term is the sum of the two previous ones. Starting instead with 1 and 3 generates the similar Lucas sequence.

$$1, 3, 4, 7, 11, 18, 29, 47, 76, \ldots$$

The 14th century Indian mathematician Narayana studied a bovine version, where female cows take 3 years to mature and produce 1 female calf each year thereafter, to produce the following sequence:

$$1, 1, 1, 2, 3, 4, 6, 9, 13, 19, 28, \ldots$$

Neighbouring terms of all these sequences, e.g. 21/13 or 76/47, increasingly approach the *golden ratio*, or φ, $\frac{1}{2}(1 + \sqrt{5})$, 1.61803399...

Can you figure out the rule which generates the Padovan sequence (*right*)?

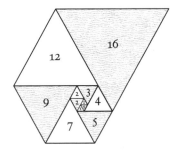

Perfect φ PHYLLOTAXIS, showing 34 and 55 spirals,
the most efficient arrangement of leaves or seeds.

Detuning (0.01% and 0.02%)
produces whorls and spokes.

ABOVE: A simulated sunflower produces a new seed every $r \times 360°$. If r is a rational fraction then the seeds line up in 'spokes', which is an inefficient packing. However, if we set $r = \frac{1}{\varphi}$ then we obtain a perfect result because no rational fraction approximates to φ. As the sunflower continues to grow, larger and larger numbers of spirals appear: $5, 8, 13, 21, 34, 55, 89, 144$ – all Fibonacci numbers of course!

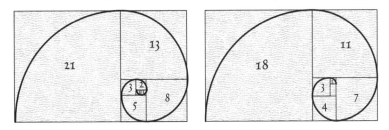

ABOVE: A rectangle formed from the Fibonacci numbers (left) and the Lucas numbers (right). As more and more squares are added, each becomes a more and more accurate golden rectangle $1:\varphi$.

π AND *e*

really irrational numbers

Pi, or π, is the ratio of the circumference of a circle to its diameter.

Archimedes [288–212 BC], seeking the value of π, started with an inscribed (internal) triangle inside a circle of diameter 1. He knew this had a perimeter of $3 \times \frac{\sqrt{3}}{2} = 2.598$ and that the perimeter of an exscribed (external) triangle was twice this, so π must be between 2.598 and 5.196. Likewise, by considering inscribed and exscribed squares, he showed that π must lie between $2\sqrt{2} = 2.828$ and 4. By considering polygons with 6, 12 and even 96 sides he narrowed his estimate to between 3.1408 and 3.1429 (*see illustration opposite*).

These days, π is calculated using a variety of series formulae. The most famous of these, first discovered by Madhava of Sangamagrama [1340–1425] is: $\pi = 4(\frac{1}{1} - \frac{1}{3} + \frac{1}{5} - \frac{1}{7} + \frac{1}{9} - \frac{1}{11} ...)$, although you will need to compute over 100 terms to do better than Archimedes!

Another unusual number is *e*, the base of natural logarithms, discovered by the Swiss mathematician Jacob Bernoulli [1655–1705] whilst investigating compound interest. In the graph of $y = e^x$ the gradient at any point is e^x and the area under the graph up to x is also e^x. This means that e^x is, uniquely, its own derivative and integral.

The value of $(1+\frac{1}{n})^n$ approaches *e* as n gets bigger and bigger (*opposite*).

π = 3.14159265358979323846264338327950288419716939937510
58209749445923078164062862089986280348253421170679...

e = 2.71828182845904523536028747135266249775724709369995
95749669676277240766303535475945713821785251664274...

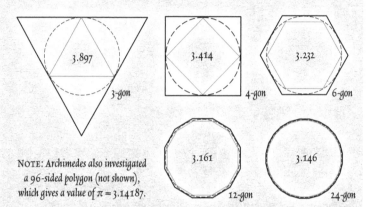

LEFT: Pi, or π, is the ratio between the circumference and the diameter of a circle. BELOW: Archimedes' method for finding the value of π $(= 3.14159\dots)$ involves drawing two similar polygons inside and outside a unit circle and finding the average between the two perimeters.

3.897   3-gon

3.414   4-gon

3.232   6-gon

3.161   12-gon

3.146   24-gon

NOTE: Archimedes also investigated a 96-sided polygon (not shown), which gives a value of $\pi = 3.14187$.

e is the limit of $(1 + \frac{1}{n})^n$ as n approaches infinity.

We use the formula for $(a + b)^n$, derived on page 42, to give:

$$(1 + \tfrac{1}{n})^n = 1 + \frac{n}{n} + \frac{n(n-1)}{2! \, n^2} + \frac{n(n-1)\,(n-2)}{3! \, n^3} + \dots$$

As n gets larger and larger, the -1's and -2's become less and less relevant, and the powers of n cancel out top and bottom to leave the series formula:

$$e = (1 + \tfrac{1}{n})^n = \frac{1}{0!} + \frac{1}{1!} + \frac{1}{2!} + \frac{1}{3!} + \frac{1}{4!} + \frac{1}{5!} + \dots = 1 + \frac{1}{1} + \frac{1}{2} + \frac{1}{6} + \frac{1}{24} + \frac{1}{120} + \dots$$

so, $e = 1 + 1 + 0.5 + 0.1666 + 0.0416 + 0.0083 + 0.0013 + \dots = 2.71828\dots$

COMPLEX NUMBERS
everything we need

The inverse of squaring or cubing involves taking a square or cube root of a number. Since squaring any number, negative or positive, produces a positive number, what can be the square root of a negative number? For centuries, the idea of taking the square root of a negative number was considered absurd. But by the 16th century a solution was needed, if mathematics was going to be complete.

Today, we define the square root of -1 as i. This new 'number' i is simply defined by the equation $i^2 = -1$. The 'i' stands for 'imaginary', a term coined by Rene Descartes in 1637. From then on, all numbers were seen as possessing a real component and an imaginary component, forming a complex number of the form $a + ib$.

It is convenient to represent complex numbers as points on a plane. The old-fashioned 'real' part of the number is represented on the x-axis and the 'imaginary' part on the y-axis. Using complex notation, we can employ all the usual rules of algebra to add, subtract, multiply and divide complex numbers, in each case replacing i^2 with -1 wherever it occurs. Importantly, with the introduction of complex numbers, no further numbers are needed for the completeness of mathematics.

MULTIPLYING COMPLEX NUMBERS

To multiply two complex numbers simply multiply out all terms and then recombine, remembering that $i^2 = -1$

E.g. $(2 + i) \times (3 + 5i)$

$= (2 \times 3) + (2 \times 5i) + (3 \times i) + (1 \times 5i^2)$

$= 6 + 10i + 3i - 5$

$= 1 + 13i$

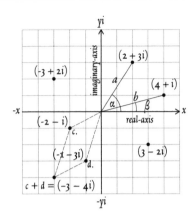

$$1/i = \frac{-i^2}{i} = -i$$

$$2^i = e^{\log(2)i} = 0.769 + 0.639\,i$$

$$i^i = e^{\log(i)i} = 0.208$$

$$\sqrt{i} = \frac{1}{\sqrt{2}} + \frac{1}{\sqrt{2}}i = 0.707 + 0.707\,i$$

$$i\sqrt{i} = e^{-\log(i)i} = 4.810$$

$$\log(i) = \frac{\pi}{2i} = 1.571\,i$$

$$\sin(i) = 1.175\,i$$

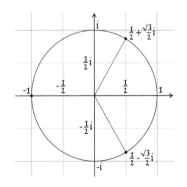

COMBINING COMPLEX NUMBERS

Two complex numbers, c and d, of the form x + yi may be added by separately adding their real parts and their imaginary parts. Any complex number x + yi can also be represented by a line with length (or modulus) a which makes an angle (or argument), α, with the x-axis. To multiply this with a second complex number, length b and angle β, you can either do the algebra (see opposite) or multiply the two moduli (ab) and add the arguments (α + β).

COMPLEX COMPLETENESS

The real numbers (i.e. all the integers, rational fractions and irrationals, including transcendentals) form a group under addition, subtraction, multiplication and division (page 17). Complex numbers form a group under all the algebraic operations including the extraction of roots. No new symbol for the cube root of -1 is needed; i is all we need. We can even assign consistent meanings to expressions such as 2^i, i^i, $\sqrt[i]{i}$, log i, sin i. Anything you can do with real numbers can be done with complex numbers too.

THE CUBE ROOTS OF -1

All the roots of 1 and -1 lie on a circle radius 1. Any complex number whose modulus is 1 and whose argument can be multiplied by 3 to get 180° is a cube root of −1. An argument of 60° obviously works, as does 180° and 270°. So −1 has three cube roots $\left(-1, \frac{1}{2} + \frac{\sqrt{3}}{2}i, \frac{1}{2} - \frac{\sqrt{3}}{2}i\right.$, shown left).

In the same way, 1 also has three cube roots. In addition to 1 itself, the other two are $-\frac{1}{2} + \frac{\sqrt{3}}{2}i$ and $-\frac{1}{2} - \frac{\sqrt{3}}{2}i$

CONTINUED FRACTIONS
a good approximation

Non-integer numbers greater than 1 can be written in a form called a continued fraction. For example: π (= 3.14159...) can be written as $3 + \frac{1}{7.0625...}$ because $\frac{1}{7.0625...} = 0.14159....$ But 7.0625... can be also be written as $7 + \frac{1}{15.996...}$ and 15.996... can be written as another fraction, etc. The result is a string of fractions over fractions (*see below*), which is a complex but perfectly valid way of representing π.

The string of coefficients 3, 7, 15, 1, 292, 1, 1... which occur in the denominators tell us a lot about the structure of the number. For example, the appearance of the relatively large number 292 in the fifth coefficient is interesting. Since $\frac{1}{292}$ is quite small, using just the first four coefficients will give us a very accurate approximation to π.

Continued fractions can distinguish rational numbers from irrational numbers. Rational numbers always have finite continued fractions (which terminate), whereas irrational numbers (like π) have unending representations. The structure of a continued fraction does not depend on the base in which it is written.

Some famous irrational numbers like the golden ratio φ (*see page 26*) and $\sqrt{2}$ (*see page 24*) have pleasingly simple patterns in their continued fractions (*see opposite*).

$$\pi = 3 + \cfrac{1}{7 + \cfrac{1}{15 + \cfrac{1}{1 + \cfrac{1}{292 + \cfrac{1}{1 + \cfrac{1}{1 + \cfrac{1}{1 + \cfrac{1}{2 + \cfrac{1}{1 + \cfrac{1}{3 + \cfrac{1}{1 + \ddots}}}}}}}}}}}$$

LEFT: *The continued fraction for π. The first four coefficients generate an approximation which is good to one part in a million.*

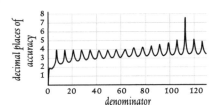

ABOVE: Accuracy of rational fraction approximations to π. Note the huge increase in accuracy when the denominator equals 113.

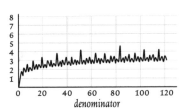

ABOVE: Accuracy of rational fraction approximations to φ. None work well because the continued fraction for φ contains only 1's.

$$\sqrt{2} = 1 + \cfrac{1}{2 + \cfrac{1}{2 + \cfrac{1}{2 + \cfrac{1}{2 + \cfrac{1}{2 + \cfrac{1}{2 + \ddots}}}}}}$$

$$\sqrt{3} = 1 + \cfrac{1}{1 + \cfrac{1}{2 + \cfrac{1}{1 + \cfrac{1}{2 + \cfrac{1}{1 + \cfrac{1}{2 + \ddots}}}}}}$$

$$\sqrt{5} = 2 + \cfrac{1}{4 + \cfrac{1}{4 + \cfrac{1}{4 + \cfrac{1}{4 + \cfrac{1}{4 + \cfrac{1}{4 + \ddots}}}}}}$$

$$\varphi = 1 + \cfrac{1}{1 + \cfrac{1}{1 + \cfrac{1}{1 + \cfrac{1}{1 + \cfrac{1}{1 + \ddots}}}}}$$

$$\pi = \cfrac{4}{1 + \cfrac{1^2}{2 + \cfrac{3^2}{2 + \cfrac{5^2}{2 + \cfrac{7^2}{2 + \cfrac{9^2}{2 + \ddots}}}}}}$$

$$e = 2 + \cfrac{1}{1 + \cfrac{1}{2 + \cfrac{1}{1 + \cfrac{1}{1 + \cfrac{1}{4 + \cfrac{1}{1 + \ddots}}}}}}$$

ABOVE: Infinite continued fractional representations of some important irrational numbers.

COMPLEMENTARY FRACTIONS
how rational is that?

Some fractions are more rational than others. Indeed, we can measure the *rationality* of a fraction using an extended form of its *complement*.

The complement of a simple fraction like $\frac{3}{5}$ is $1 - \frac{3}{5} = \frac{2}{5}$, the two summing to 1. However, the complement of a vulgar fraction is more complex. For example, a fraction like $\frac{67}{18}$ can be written as $3 + \frac{13}{18}$ or, choosing the nearest integer, as $4 - \frac{5}{18}$. To find the complement, add 1 to the whole number if the sign is positive or subtract 1 if the sign is negative, and then reverse the sign. This produces the fractions $4 - \frac{13}{18}$ or $3 + \frac{5}{18}$, equal to $\frac{59}{18}$. Note that adding a fraction to its complement always produces a whole number, so here $\frac{67}{18} + \frac{59}{18} = \frac{126}{18} = 7$.

Fractions can be efficiently broken down into a series of smaller fractional terms in a *nearest integer continued fraction*, e.g.:

$$\frac{67}{18} = 4 - \cfrac{1}{3 + \cfrac{1}{2 - \frac{1}{3}}}$$

The *first complement* of this fraction involved changing the 4 into a 3 and the first minus into a plus. To generate the *second complement* we move to the second row, and change the 3 into a 4 and the plus into a minus; to generate the *third complement* we move to the third row and change the 2 into a 1 and the last minus into a plus. Applying this process to every term in a continued fraction produces the *full complement* of the fraction. In the case of $\frac{67}{18}$ the full complement is:

$$3 + \cfrac{1}{4 - \cfrac{1}{1 + \frac{1}{3}}} = \frac{43}{13}$$

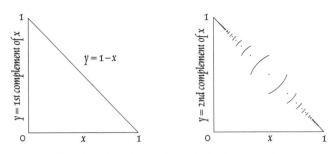

Above: The first and second complements of all the fractions between 0 and 1. The graph is discontinuous at $\frac{1}{2}$, $\frac{1}{3}$, $\frac{1}{4}$, $\frac{1}{5}$, etc, and their complements, $\frac{1}{2}$, $\frac{2}{3}$, $\frac{3}{4}$, $\frac{4}{5}$.

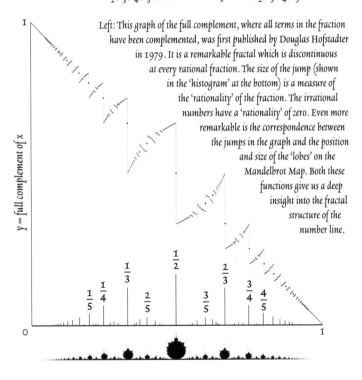

Left: This graph of the full complement, where all terms in the fraction have been complemented, was first published by Douglas Hofstadter in 1979. It is a remarkable fractal which is discontinuous at every rational fraction. The size of the jump (shown in the 'histogram' at the bottom) is a measure of the 'rationality' of the fraction. The irrational numbers have a 'rationality' of zero. Even more remarkable is the correspondence between the jumps in the graph and the position and size of the 'lobes' on the Mandelbrot Map. Both these functions give us a deep insight into the fractal structure of the number line.

FACTORIAL NUMBERS
permutations and combinations

How many ways are there of shuffling a deck of 52 cards? The top card can be any one of 52, the second any one of 51, and so on. So the total number of different ways of ordering 52 cards is 52 × 51 × 50 × 49 × ... × 3 × 2 × 1. This number is the factorial of 52 and is written 52! It is a prodigiously large number, approximately equal to 10^{68} (or 10 with 68 zeros after it). So there are $n!$ ways of shuffling (*permuting*) n cards.

How many ways are there of choosing a hand of 13 cards from a deck of 52? As before, there are 52 possibilities for the first card, 51 for the second, etc; so the total number of ways is 52 × 51 × 50 × ... × 42 × 41 × 40 (stopping after you have chosen 13 cards), which can be written more succinctly as $\frac{52!}{39!}$. Actually, this overstates the case, as the order in which they were dealt to you is irrelevant and you will probably sort your hand sort into suits anyway. As we have seen, there are $n!$ ways of ordering n cards, so we must divide the high figure by 13! to get the right answer, which is 635,013,559,600 or a little over a trillion different hands.

We can now write down a general formula for the number of different ways k objects can be chosen from a set of n objects. This function is called a combination and is written $_nC_k$ or $\binom{n}{k}$

$$= \frac{n!}{(n-k)! \times k!}$$

$0! = 1$
$1! = 1$
$1 \times 2 = 2$
$1 \times 2 \times 3 = 6$
$1 \times 2 \times 3 \times 4 = 24$
$1 \times 2 \times 3 \times 4 \times 5 = 120$
$1 \times 2 \times 3 \times 4 \times 5 \times 6 = 720$
$1 \times 2 \times 3 \times 4 \times 5 \times 6 \times 7 = 5,040$
$1 \times 2 \times 3 \times 4 \times 5 \times 6 \times 7 \times 8 = 40,320$
$1 \times 2 \times 3 \times 4 \times 5 \times 6 \times 7 \times 8 \times 9 = 362,880$
$1 \times 2 \times 3 \times 4 \times 5 \times 6 \times 7 \times 8 \times 9 \times 10 = 3,628,800$
$1 \times 2 \times 3 \times 4 \times 5 \times 6 \times 7 \times 8 \times 9 \times 10 \times 11 = 39,916,800$
$1 \times 2 \times 3 \times 4 \times 5 \times 6 \times 7 \times 8 \times 9 \times 10 \times 11 \times 12 = 479,001,600$

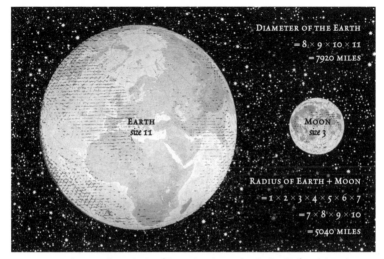

DIAMETER OF THE EARTH
$= 8 \times 9 \times 10 \times 11$
$= 7920$ MILES

EARTH
size 11

MOON
size 3

RADIUS OF EARTH + MOON
$= 1 \times 2 \times 3 \times 4 \times 5 \times 6 \times 7$
$= 7 \times 8 \times 9 \times 10$
$= 5040$ MILES

ABOVE: Ancient Metrology. The size of the Earth and Moon in miles involve factorial numbers.

ABOVE: The number of different ways of ordering a complete deck of 52 cards is around the same as the number of atoms in our galaxy:

$52 \times 51 \times 50 \times 49 \times 48 \times 47 \times 46 \times 45$
$\times 44 \times 43 \times 42 \times 41 \times 40 \times 39 \times 38 \times$
$37 \times 36 \times 35 \times 34 \times 33 \times 32 \times 31 \times 30 \times$
$29 \times 28 \times 27 \times 26 \times 25 \times 24 \times 23 \times 22$
$\times 21 \times 20 \times 19 \times 18 \times 17 \times 16 \times 15 \times$
$14 \times 13 \times 12 \times 11 \times 10 \times 9 \times 8 \times 7 \times 6$
$\times 5 \times 4 \times 3 \times 2 \times 1 = 52!$

ABOVE: The no. of ways of picking 13 cards is

$$52 \times 51 \times 50 \times 49 \times 48 \times 47 \times 46 \times 45 \times 44 \times 43 \times 42 \times 41 \times 40 = \frac{52!}{39!}$$

The number of ways of then ordering these 13 cards is 13! So the total number of possible bridge hands of 13 cards is …

$$\frac{52!}{39! \times 13!}$$

ALGEBRAIC NUMBERS
the familiar family

The decimal expansion of an irrational number is infinitely long without ever repeating. Numbers like $\sqrt{2}$ *(page 24)*, φ *(the golden ratio, page 26)*, *e (the base of natural logarithms)* and π *(page 28)* are all irrational.

All numbers which can be expressed in terms of the algebraic operations of addition, subtraction, multiplication, division, raising to the power and taking roots are called *algebraic numbers*. Thus, $\sqrt{2}$, a solution to the equation $x^2 = 2$, and φ, a solution to the equation $x^2 = x + 1$, are algebraic, while *e* and π are not.

All rational numbers are algebraic, and we saw earlier how each corresponds to a unique integer *(page 23)*. But can we now add the algebraic irrationals and still preserve the one-to-one correspondence with the integers? Amazingly, there is a way of coding every algebraic number into an ordered string of positive integers which can be turned into a unique integer *(shown opposite)*. This proves that the cardinality of the algebraic irrationals is the same as that of the integers.

It looks as if the number line is pretty full now. Between all the integers we have put an infinite number of rational fractions, and somehow we have squeezed in all the algebraic irrationals as well, without increasing the cardinality of the line.

Surely there can't be any more numbers on the line, can there?

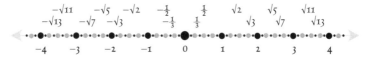

To encode an Algebraic Number x into an Integer:

Example: $\sqrt{5}$

1. Write down an equation which defines x.

$x^2 = 5$

2. By standard algebraic operations, turn the equation into standard form (with ordered integer coefficients).

$1x^2 + 0x - 5 = 0$

3. Write down the coefficients in order.

$1, 0, -5$

4. Double all the positive coefficients, Make all the negative coefficients positive, then double them and subtract 1.

$2, 0, 9$

5. The number we want to code ($\sqrt{5}$) is actually the second solution to this equation, the first being $-\sqrt{5}$. So add the number 1 to the list (starting counting from zero.)

$2, 0, 9, 1$

6. Working from the right, calculate the number obtained by raising all the primes in sequence to these exponents. Multiply them all together. This is the unique code for $\sqrt{5}$.

$7^2 \times 5^0 \times 3^9 \times 2^1$
$= 1928934$

7. You can reverse the process to extract the original number from its code, e.g. try finding the number whose code is 1470. Familiar?

1470
$= 7^2 \times 5^1 \times 3^1 \times 2^1$

Angle (degrees)	0	30°	45°	60°	90°	180°	270°	360°
Angle (radians)	0	$\pi/6$	$\pi/4$	$\pi/3$	$\pi/2$	π	$3\pi/2$	2π
sin	0	$1/2$	$1/\sqrt{2}$	$\sqrt{3}/2$	1	0	-1	0
cos	1	$\sqrt{3}/2$	$1/\sqrt{2}$	$1/2$	0	-1	0	1
tan	0	$1/\sqrt{3}$	1	$\sqrt{3}$	∞	0	∞	0
cosec = 1 / sin	∞	2	$\sqrt{2}$	$2/\sqrt{3}$	1	∞	-1	∞
sec = 1 / cos	1	$2/\sqrt{3}$	$\sqrt{2}$	2	∞	-1	∞	1
cotan = 1 / tan	∞	$\sqrt{3}$	1	$1/\sqrt{3}$	0	∞	0	∞

ABOVE: TRIGONOMETRIC NUMBERS are algebraic irrational numbers derived from the sines, cosines and tangents of angles which divide a circle into an integer number of segments.

TRANSCENDENTAL NUMBERS
and transfinite ones too

In 1844, French mathematician Joseph Liouville [1809–1882] proved that you could construct *transcendental numbers*, irrational numbers which were not algebraic. In 1873, his compatriot Charles Hermite [1822–1901] proved that the base of natural logarithms $e = 2.7182818285...$ was transcendental, and a few years later π. In 1874, German mathematician Georg Cantor [1845–1918] showed that the transcendentals are uncountable (*opposite*), i.e. there are more points on a line between 0 and 1 than there are integers, even though the number of integers is infinite. Cantor proposed that the cardinality of the integers should be called $\aleph_0$, "aleph-nought", and that the cardinality of the irrationals (including the transcendentals) should be called $\aleph_1$, "aleph-one".

We have seen that $\infty \times anything = \infty$ (e.g. the number of integers is equal to the number of even numbers). Even $\infty \times \infty = \infty$ (e.g. the number of ordered pairs or rational fractions is equal to the number of integers, *see page xx*). But we have not tried 2^∞.

There are 2^n ways of selecting any number of objects from a pile of n objects (*see page xx*) and each of these ways can be associated with a binary number, e.g., the binary number 101 could stand for 'pick the 1st and 3rd objects'. For an infinite pile, the 2^∞ ways must be equal to the number of ways of writing an irrational binary fraction, and since this is effectively all the irrationals, Cantor concluded that $2^\infty = \aleph_1$.

Cantor did not stop there. He reasoned that if the cardinality of the irrationals was somehow greater than the cardinality of the integers, then a whole series of *transfinite numbers* could also be constructed. So if $2^\infty = \aleph_1$, then $2^{\aleph_1} = \aleph_2$ and $2^{\aleph_2} = \aleph_3$ ... and so on.

0.110001000000000000000000010000...

ABOVE: The 'LIOUVILLE CONSTANT': The ones appear in the 1st, 2nd, 6th, 24th, etc decimal places, where the sequence 1, 2, 6, 24... is the factorial sequence 1!, 2!, 3!, 4! etc. This bizarrely constructed number can never be the solution to a finite polynomial equation—it is a 'transcendental number'.

1	0.[2]746238184...	
2	0.3[7]82352448...	
3	0.15[6]3057435...	
4	0.036[2]454883...	
5	0.5283[6]47283...	
6	0.15399[5]6238...	
7	0.274527[3]544...	
8	0.2476554[0]47...	
9	0.56394527[9]7...	
10	0.936459476[4]...	

CANTOR'S DIAGONAL ARGUMENT

Suppose the irrational numbers were countable—i.e. there was some way to pair up every irrational number with at least one integer. The start of the list might look like as shown (left).

Now pick out a diagonal line of digits in bold which can be thought of as an irrational number 0.2762653094...

If we now add 1 (mod 10) to every digit, we can immediately see that the number 0.3873764105... cannot be in the list because it differs from the 1st number in the 1st digit, the 2nd number in the 2nd digit, the 3rd number in the 3rd digit, etc.

The inescapable conclusion is that our supposedly complete list of irrationals is not complete and cannot be counted.

NATURAL—Start with the counting numbers (zero may be included).

Cardinality
$$N \quad \aleph_0$$

INTEGER—Extend the line backward to include the negatives.

$$Z \quad \aleph_0$$

RATIONAL—Insert all the fractions.

$$Q \quad \aleph_0$$

REAL ALGEBRAIC—Insert all the roots.

$$A_r \quad \aleph_0$$

REAL—Add the transcendental numbers to make a continuous line.

$$R \quad \aleph_1$$

41

PASCAL'S TRIANGLE
and the binomial theorem

Pascal's famous triangle (*opposite*) is an arrangement of numbers where each number is the sum of the two above it. It famously generates the coefficients of the expansion of the binomial $(a + b)^n$ and contains the triangular numbers and the Fibonacci numbers (*see opposite*).

It also lists all the possible combinations of objects selected out of a pile of n objects (*lower, opposite*). The k coefficient in the nth line is the number of ways of choosing k objects from n objects (starting counting from zero), or $_kC_n$.

We can go further and say that the kth coefficient in the expansion of $(a + b)^n$ is also $_kC_n$. This important result is known as the *binomial theorem* and can be stated as follows:

$$(a + b)^n = a^n + \frac{n}{1!} a^{(n-1)}b + \frac{n(n-1)}{2!} a^{(n-2)}b^2 + \frac{n(n-1)(n-2)}{3!} a^{(n-3)}b^3 + \cdots$$

A simpler form has $a = 1$:

$$(1 + x)^n = 1 + \frac{n}{1!} x + \frac{n(n-1)}{2!} x^2 + \frac{n(n-1)(n-2)}{3!} x^3 + \cdots$$

or,

$$(x + y)^n = \sum_{k=0}^{n} \binom{n}{k} x^{n-k} y^k$$

Example expansions are shown in the illustration (*centre opposite*), clearly showing how Pascal's triangle supplies the coefficient numbers for each line.

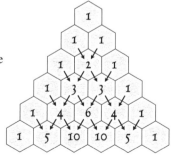

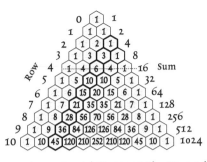
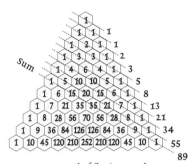

ABOVE: PASCAL'S TRIANGLE. *The uppermost digit represents row zero. The following rows then display their row number as their second digit. The numbers in each row, n, sum to 2^n. The third number in the row is the nth triangular number and the fourth is the tetrahedral number (see page 10). Diagonals display the triangular and tetrahedral numbers, and sum to produce the Fibonacci numbers.*

RIGHT: *The numbers in Pascal's triangle produce the coefficients of binomial power expansions.*

EXAMPLE: *the expansion of $(a+b)^4$ uses the 4th line of the triangle to give the 1-4-6-4-1 pattern in the terms.*

$$a^4 + 4a^3b + 6a^2b^2 + 4ab^3 + b^4$$

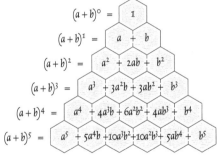

$$(a + b)^0 = 1$$
$$(a + b)^1 = a + b$$
$$(a + b)^2 = a^2 + 2ab + b^2$$
$$(a + b)^3 = a^3 + 3a^2b + 3ab^2 + b^3$$
$$(a + b)^4 = a^4 + 4a^3b + 6a^2b^2 + 4ab^3 + b^4$$
$$(a + b)^5 = a^5 + 5a^4b + 10a^3b^2 + 10a^2b^3 + 5ab^4 + b^5$$

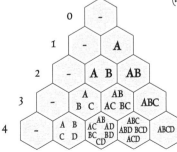

LEFT: POWER SETS. *How many different ways are there of choosing from a pile of n objects? The answer, the power set of the collection, is 2^n, the nth line of Pascal's triangle, because for each object, either you choose it or you don't, so there are the same number of possible choices of n objects as there are n-digit binary numbers.*

43

NUMBER SERIES

sequences, differences and sums

Any list of numbers generated by a rule forms a sequence, e.g. 1, 2, 3,... The most simple sequences are arithmetic or geometric (*see opposite*).

You can tell a lot about a sequence by repeatedly calculating the differences between consecutive terms, as shown for the sequence of square and tetrahedral numbers (*below*). For example, any polynomial sequence, generated by a formula such as $a + bn + cn^2 + ...,$ will reduce to a line of zeros and from the number of lines you can deduce the order of the polynomial. In fact, every polynomial generates a unique series of initial numbers, N_0, N_1, N_2, ... (shown highlighted for the square sequence), and in 1687 Isaac Newton published a formula from which, given just these numbers, you could reconstruct the whole polynomial:

$$y = N_0 + N_1 x + \frac{1}{2!}N_2 x(x-1) + \frac{1}{3!}N_3 x(x-1)(x-2) + \cdots$$

If you put $N_0 = 1$, $N_1 = 3$, $N_2 = 2$ and $N_3 = 0$ (*the numbers highlighted below, left*) and multiply out all the brackets you will end up with the equation $y = (1+x)^2$. The reason why you get $(1+x)^2$ and not x^2 is that, as a good mathematician, Newton starts counting at zero.

Not all sequences work out like this. Take the geometrical sequence 1, 2, 4, 8, 16, 32, ... The next line of differences is 1, 2, 4, 8, 16,... Can you see why?

1	4	9	25	36	49	64	81	...
3	5	7	9	11	13	15	...	
2	2	2	2	2	2	...		
0	0	0	0	0	...			

Successive differences for the square sequence

1	4	10	20	35	56	84	120	...
3	6	10	15	21	28	36	...	
3	4	5	6	7	8	...		
1	1	1	1	1	...			
0	0	0	0	...				

Differences for the tetrahedral sequence

ARITHMETIC SEQUENCES

take the form $a_n = a_1 + d(n-1)$
where a_1 is the first term
and d is the common difference.

e.g. with $a_1 = 5$ and $d = 3$,
we obtain
5, 8, 11, 14, 17, ...
+3 +3 +3 +3

the corresponding ARITHMETIC SERIES
$5 + 8 + 11 + 14 + 17 + ...$

has the SUM to the nth term of
$$\frac{n}{2}(a_1 + a_n) = \frac{n}{2}(2a_1 + (n-1)d)$$

SUM S OF AN ARITHMETIC SERIES WHICH STARTS WITH d

$S = d + 2d + 3d + ... + (n-2)d + (n-1)d + nd$
Reversing this we get
$S = n + (n-1)d + (n-2)d + ... + 3d + 2d + d$
Adding the two together,
$2S = (1+n)d + (1+n)d + ... + (1+n)d + (1+n)d$
$= n(1+n)d$
So $S = \frac{n}{2}(1+n)d$

This is the proof which Gauss
invented when he was 8 years old
(see page 8) though no doubt it was
known to the Greeks. It implies that
the sum of an arithmetic sequence
increases as the square of n.

GEOMETRIC SEQUENCES

take the form $a_n = a_1 \times r^{(n-1)}$
where a_1 is the first term
and r is the common ratio.

e.g. with $a_1 = 3$ and $r = 2$,
we obtain
3, 6, 12, 24, 48, ...
×2 ×2 ×2 ×2

the corresponding GEOMETRIC SERIES
$3 + 6 + 12 + 24 + 48 + ...$
has the SUM to the nth term of
$$\frac{a_1(1 - r^n)}{(1 - r^n)}$$

SUM S OF A GEOMETRIC SERIES WHICH STARTS WITH 1.

$S = 1 + r + r^2 + ... + r^n$
Multiply both sides by r
$rS = r + r^2 + ... + r^n + r^{n+1}$
and subtract S from both sides
$rS - S = S(r-1) = r^{n+1} - 1$

$$S = \frac{r^{n+1} - 1}{r - 1}$$

One interesting consequence of this
proof is that any number of the
form $a^n - 1$ will be divisible by $a - 1$.
E.g. $7^3 - 1$ (= 342) must be divisible
by 6, and any number of the form
$10^n - 1$ must be divisible by 9.

EUCLID'S ALGORITHM
finding the highest common factor

The *Euclidean algorithm* appears in Euclid's *Elements* [c.300 BC], and is one of the oldest algorithms in common use. Euclid may have obtained it from Eudoxus of Cnidus [408–355 BC]. Many centuries later, the algorithm was independently discovered in both India and China.

Suppose you want to find the *highest common factor* (HCF), N, of the numbers 735 and 546. We can say that $pN = 735$ and $qN = 546$, where p and q are integers.

What can we say about the difference between the two numbers, i.e., $735 - 546$? Well, this will be $pN - qN = (p-q)N = 189$. It is immediately clear that $(p-q)$ is another integer which, multiplied by the HCF, produces the difference. In fact, we can subtract any number of copies of the smaller number from the larger, and the result will still be divisible by the HCF. If we go on subtracting multiples of the smaller from the larger until there is nothing left to do, the only number left will be the HCF (*see example below*).

THE EUCLIDEAN ALGORITHM

ACTION	A	B
	546	735
Subtract 546 from B	546	189
Subtract 2 x 189 = 378 from A	168	189
Subtract 168 from B	168	21
Subtract 8 x 21 = 168	0	21

So, the highest common factor of 546 and 735 is 21

THREE IMPORTANT THEOREMS

HCF THEOREM 1:
If a is not divisible by some prime p, then
the highest common factor of a and p is 1.   HCF $(a, p) = 1$

A prime number has no factors other than itself and 1.
So if a is not divisible by p
then it can't be divisible by any factor of p either
because p has no factors (other than 1).

HCF THEOREM 2:
If a and b are coprime, then HCF $(a, b) = 1$
and the only factor which na and nb have in common is n.

Two numbers are coprime if they have no factors in common.
e.g. 21 and 25 are coprime.
The HCF of 21 x 3 = 63 and 25 x 3 = 75 must therefore be 3.

HCF THEOREM 3:
If a and b are both divisible by n,
then HCF (a, b) is also divisible by n.

The HCF contains all the factors common to a and b.
So if a and b share a factor n,
then the HCF (a, b) must also share the same factor.
e.g. we know that HCF (546, 735) is 21.
Now both 546 and 735 are divisible by 7.
So the HCF must also be divisible by 7.

THE FUNDAMENTAL THEOREM
of arithmetic

Sometimes the most self-evident statements in mathematics are the most difficult to prove. For example, we know that any number can be expressed as a product of its *prime factors*, e.g. $84 = 2 \times 2 \times 3 \times 7$, but is this factorisation *unique*? How can you be sure that there isn't some large number which is the product of primes in two different ways?

The uniqueness of prime factorisation is called the *Fundamental Theorem of Arithmetic* (FTA). To prove it, we first we need to prove *Euclid's lemma* (*below*), before the rest follows (*opposite*).

EUCLID'S LEMMA

If $a \times b$ is divisible by some prime p
then either a is divisible by p or b is divisible by p (or both).

PROOF

If a is divisible by p the lemma is proved, so let a be indivisible by p

By HCF theorem 1, HCF $(a, p) = 1$ (since p is prime), and
By HCF theorem 2, HCF $(ab, pb) = b$.

Since $a \times b$ is divisible by p, then $ab = np$ (where n is an integer).
So, HCF $(np, pb) = b$.

By HCF theorem 3, since both np and pb are divisible by p,
b must be divisible by p (but only if p is prime!). Q.E.D.!

Example, if 210 (= 14×15) is divisible by 7,
then either 14 or 15 (or both) are divisible by 7.

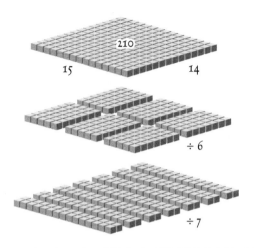

210 (which is equal to 15 × 14) can be divided by both 6 and 7.

A 15 × 14 rectangle of 210 blocks can be divided into 6 groups by cutting the rectangle horizontally and vertically.

However, the only way to cut the rectangle into 7 groups (a prime number) is by cutting it horizontally, demonstrating that at least one of the numbers 15 and 14 must be divisible by 7.

÷ 6

÷ 7

THE FUNDAMENTAL THEOREM OF ARITHMETIC

The Prime Factorisation of Any Number is Unique

PROOF

Suppose that a certain number $N = p_1 \times p_2 \times p_3 \ldots = q_1 \times q_2 \times q_3 \ldots$
where p_n and q_n are prime with all the p's different from the q's.
(If any are the same, divide by that prime to obtain a new N.)

Since N is divisible by p_1 then by Euclid's lemma,
q_1 or q_2 or $q_3 \ldots$ must be divisible by p_1).

But we know p_1 cannot divide q_1 or q_2 or $q_3 \ldots$
because the p's and q's are different primes.

CONCLUSION: it is impossible to express any number N
as a product of different primes.
The original factorisation must be unique. Q.E.D.!

FERMAT'S LITTLE THEOREM
and the pseudoprimes

Fermat's Little Theorem states that if p is a prime number and a is any integer, then $a^p - a$ will be an integer multiple of p. In modular notation (*see page 20*), this is expressed as $a^p \equiv a \pmod{p}$.

Most simply, when $a = 2$, for the prime number p, $2^p - 2$ will be divisible by p, e.g. $2^7 - 2 = 126 = 7 \times 18$. So, $2^p \equiv 2 \pmod{p}$.

We can turn this theorem round and say that if 2^p divided by p gives a remainder 2, then there is a good chance that p is prime. In fact, a number which passes Fermat's test but which is not prime is called a *pseudoprime*. Pseudoprimes are rare, so the test is a good one. Indeed, there are 171 numbers under 1000 which pass Fermat's test, of which only 3 (341, 561 and 645) are not prime (341 is divisible by 11 and the other two are divisible by 3). All in all, 455,067,395 numbers under 10 billion satisfy Fermat's test, of which only 14,884 (0.003%) are pseudoprimes. When searching for really large primes, such as those used to make secure online payments, it is much faster to stick to numbers which satisfy Fermat's criterion, as there are simple ways to calculate the remainder of $\frac{2^p}{p}$ without calculating 2^p (*see opposite*).

We can visualise this theorem using stepping stones (*see page 20*). Instead of moving round using a fixed number, we instead multiply the number on the stone by a jump factor a (*shown opposite*). If the number of stones N is prime, then it is impossible to stand still on any Stone R (except Stone zero) and all paths home have the same number of jumps. If Stone 1 returns to Stone 1 after $N-1$ jumps then one more jump will bring us to Stone a. This implies that if N is prime then $a^N \equiv a \pmod{N}$, or $a^N - a$ is divisible by N, Fermat's Little Theorem in its general form.

FERMAT'S LITTLE THEOREM

For any integer a and prime p:

$$a^{p-1} \equiv 1 \pmod{p}, \text{ or } a^p \equiv a \pmod{p}$$

When $a = 2$, then $2^{p-1} \equiv 1 \pmod{p}$, or $2^p \equiv 2 \pmod{p}$

TO CALCULATE $2^p \pmod{p}$: ✱ Write p in binary. ✱ Let $a = 1$. ✱ Use the digits of p, left to right. ✱ For each digit, square a, and if the digit is odd then double a. ✱ Replace a with $a \pmod{p}$ and repeat. ✱ When you run out of digits, $a \equiv 2^p \pmod{p}$.

EXAMPLE: ✱ Calculate $2^{11} \pmod{11}$. ✱ Write 11 in binary = 1 0 1 1. ✱ Let $a = 1$. ✱ Work on 1-0-1-1 left to right. ✱ The first digit is 1, so square and double a, (i.e. $a = 2$). ✱ The next digit is zero so square a, (i.e. $a = 4$). ✱ The next digit is 1 so square and double a, (i.e. $a = 2 \times 4^2 = 32$). ✱ Replace a with $a \pmod{p}$, i.e. $a = 32 \equiv 10 \pmod{11}$. ✱ Square, double and mod one last time, i.e. $a = 200 \equiv 2 \pmod{11}$.
So, $2^{11} \pmod{11} \equiv 2$. Or $2048 = 186 \times 11 + 2$

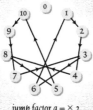

jump factor $a = \times 2$

jump factor $a = \times 3$

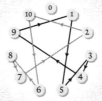

jump factor $a = \times 5$

jump factor $a = \times 7$

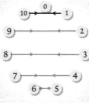

jump factor $a = \times 11$

Stepping stone paths for N = 11. Starting at Stone 1, every path returns to Stone 1 after 10 jumps, even those with a cyclic period of 2 or 5. This means that a^{10} leaves remainder 1 when divided by 11, and a^{11} will leave remainder a when divided by 11.

MERSENNE PRIMES
and perfect numbers

Mersenne numbers are of the form $2^n - 1$. They were discovered by the French monk Marin Mersenne [1588–1648]. Many are prime, but if n is a composite number the Mersenne number cannot be prime because $2^{ab} - 1$ is always divisible by both $2^a - 1$ and $2^b - 1$ (*see opposite*).

On the other hand, if n is prime, then there is at least a possibility that $2^n - 1$ is prime. This works for $n = 2$, 3, 5 and 7 but not for $n = 11$. In fact, although 12 of the first 32 prime numbers generate Mersenne primes, after that they get exceedingly rare. Currently, 51 Mersenne primes are known, the largest of which is $2^{74,207,281} - 1$, also the largest known prime. Nobody knows whether or not the number of Mersenne primes is infinite.

There is an interesting relationship between the Mersenne primes and perfect numbers (*shown below*). For every Mersenne prime M_p there is a perfect number $P = \frac{1}{2}M_p(M_p + 1)$. This equation generates all known perfect numbers and shows that they are triangular numbers. When written in binary, all perfect numbers have a string of p ones followed by a string of $(p-1)$ zeros, e.g. 28 = 11100, and 8128 = 1111111000000.

PRIME	MERSENNE PRIMES	PERFECT NUMBERS
p	$M_p = 2^p - 1$	$P = \frac{1}{2}M_p(M_p + 1) = M_p \times 2^{p-1}$
2	3	$\frac{1}{2}(3 \times 4) = 3 \times 2^1 = 6$
3	7	$\frac{1}{2}(7 \times 8) = 7 \times 2^2 = 28$
5	31	$\frac{1}{2}(31 \times 32) = 31 \times 2^4 = 496$
7	127	$\frac{1}{2}(127 \times 128) = 127 \times 2^6 = 8128$
13	8191	$\frac{1}{2}(8191 \times 8192) = 8191 \times 2^{12} = 33550336$

PROOF THAT $2^{ab} - 1$ IS ALWAYS COMPOSITE

* 2^n in binary is one followed by n zeros.
* Subtract one from this number and you get a string of n ones.
* If n is a composite number, i.e. $n = ab$, then you can always split the n ones into a groups of b (or b groups of a).

EXAMPLE: $2^{15} - 1$ in binary is 111 111 111 111 111 (which splits into 5 groups of 111). But this number can also be generated by multiplying 111 by 100100100100 because every 1 in the multiplicand turns into 111 with nothing to carry. Therefore $2^{15} - 1$ must be divisible by 7 (111 in binary).

Can you also see why all numbers of the form $a^n - 1$ are never prime when $a > 2$?

hint: see page 51

n	MERSENNE NO.		STATUS	n	MERSENNE PRIME
1	$2^1 - 1$	1	–	2	3
2*	$2^2 - 1$	3	prime	3	7
3*	$2^3 - 1$	7	prime	5	31
4	$2^4 - 1$	15	composite	7	127
5*	$2^5 - 1$	31	prime	13	8191
6	$2^6 - 1$	63	composite	17	132071
7*	$2^7 - 1$	127	prime	19	524287
8	$2^8 - 1$	255	composite	31	2147483647
9	$2^9 - 1$	511	composite	61	2305843009213693951
10	$2^{10} - 1$	1023	composite	89	27 digits
11*	$2^{11} - 1$	2047	composite	107	33 digits
12	$2^{12} - 1$	4095	composite	127	39 digits
13*	$2^{13} - 1$	8191	prime	521	157 digits

ABOVE: A short table of Mersenne numbers. Prime exponents are starred but not all such primes generate Mersenne primes. For example, to see why 11 does not work see page 51.

HYPERCOMPLEX NUMBERS
into the next dimension

The number plane is formed by an R-axis of real numbers, and a perpendicular *i*-axis of imaginary numbers (*see page 30*). If we extend the system into a third dimension using a second imaginary *j*-axis, we find that addition works, but $i \times j$ runs into inconsistencies. In 1843, the Irish mathematician William Rowan Hamilton [1805–1865] realised that to be able to reliably multiply *i* by *j* you need to add one more imaginary dimension *k*, with multiplication rules as shown opposite. The resulting four-dimensional *quaternion* number system is useful for describing rotations in three dimensions, and so has many applications.

Going even higher, eight-dimensional *octonions* can be used to describe rotations in seven-dimensional space, and have been proposed as a unifying framework for subatomic physics.

In the same way that complex numbers are the basis for the two-dimensional Mandelbrot set (*below left*), these hypercomplex numbers can be used to draw some amazing three-dimensional fractals (*e.g. right, by Daniel White*).

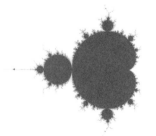

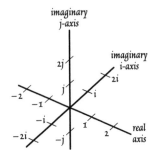

LEFT: Adding a second imaginary axis to the complex plane produces a three-dimensional 'ternion', of the form $a + bi + cj$

Adding ternions is easy e.g.

$$(1 + 2i) + (i + 3j) = 1 + 3i + 3j$$

however multiplying ternions involves defining $i \times j$, which quickly produces inconsistencies.

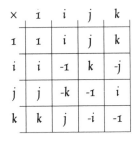

×	1	i	j	k
1	1	i	j	k
i	i	-1	k	-j
j	j	-k	-1	i
k	k	j	-i	-1

LEFT: Quarternions are 4-dimensional. The four elements multiply as shown, so

$$i \times j = k \qquad j \times k = i \qquad k \times i = j$$

Real number multiplication is commutative, i.e. 3×2 is the same as 2×3. However, quaternion multiplications are not commutative (neither are rotations in 3-D, below), so

$$j \times i = -k \qquad k \times j = -i \qquad i \times k = -j$$

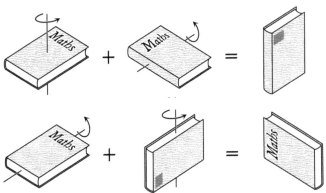

ABOVE: Like quaternion multiplications, rotations in 3-D are non-commutative. The same moves in a different order produce a different result.

MYSTERIOUS NUMBERS
the dimensionless physical constants

The *dimensionless constants* are a group of numbers which define the world we live in. They are 'dimensionless' because they take the same value whatever system of units are used to calculate them.

Take, for example, the ratio of the mass of an electron to that of a proton. The best measurement of this is 1836.15267343, but nobody has a clue why it has this value. We only know that if this and the other ratios shown opposite were different, then the entire universe would also be very different, probably with no humans in it.

Both the electromagnetic and the gravitational force obey an inverse square law. This means that the ratio of the strengths of these two forces is independent of distance, and is equal to the dimensionless number $e^2/4\pi\varepsilon_0 Gm_e^2$ or 4.17×10^{42}. In other words, the electrostatic force of repulsion between two electrons is 4 million trillion trillion trillion times greater than their gravitational attraction!

Any two points in space separated by a distance r have a quantum force F_q associated with them, equal to $= hc/2\pi r^2$. The ratio of this quantum force to the electromagnetic and gravitational forces is particularly significant. The electromagnetic force is 137 times weaker than the quantum force and the gravitational force is 5.7×10^{44} times weaker.

Could these constants be mathematical in nature? No-one knows.

MASS *of an* ELECTRON: $m_e = 9.1094 \times 10^{-31}$ kg   SPEED *of* LIGHT: $c = 2.998 \times 10^8$ ms^{-1}

MASS *of a* PROTON: $m_p = 1.6726 \times 10^{-27}$ kg   PLANCK's *constant*: $h = 6.6261 \times 10^{-34}$ J s

MASS *of a* NEUTRON: $m_n = 1.6749 \times 10^{-27}$ kg   PERMITTIVITY *const*.: $\varepsilon_0 = 8.8542 \times 10^{-12}$ F m^{-1}

CHARGE *on an* ELECTRON: $e = 1.6022 \times 10^{-19}$ C   GRAVITATION *const*.: $G = 6.6743 \times 10^{-11}$ N m^2 kg^{-2}

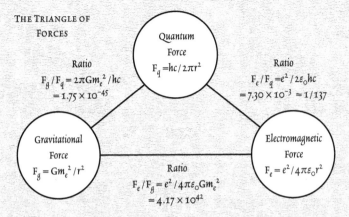

Quantum
Force
$F_q = hc / 2\pi r^2$

Ratio
$F_g / F_q = 2\pi Gm_e^2 / hc$
$= 1.75 \times 10^{-45}$

Ratio
$F_e / F_q = e^2 / 2\varepsilon_0 hc$
$= 7.30 \times 10^{-3} \approx 1/137$

Gravitational
Force
$F_g = Gm_e^2 / r^2$

Electromagnetic
Force
$F_e = e^2 / 4\pi\varepsilon_0 r^2$

Ratio
$F_e / F_g = e^2 / 4\pi\varepsilon_0 Gm_e^2$
$= 4.17 \times 10^{42}$

The ratio $F_e / F_g = e^2 / 2\varepsilon_0 hc = 0.007297 \approx 1/137$, above, is the important Fine Structure Constant, which governs the strength of the electromagnetic interaction between charged particles. The equivalent number for gravity $F_g / F_q = 2\pi Gm_e^2 / hc = 1.75 \times 10^{-45}$ is not so well known, but it may have an important role to play in a quantum theory of gravity. As with the electron:proton mass ratio, below, no one has a clue why these constants are the size they are. Is there some deep mathematical inevitability about them? Or not?

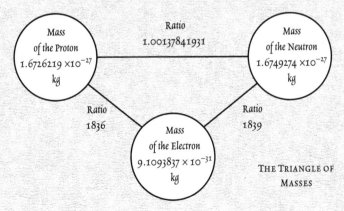

Ratio
1.00137841931

Mass
of the Proton
$1.6726219 \times 10^{-27}$
kg

Mass
of the Neutron
$1.6749274 \times 10^{-27}$
kg

Ratio
1836

Ratio
1839

Mass
of the Electron
$9.1093837 \times 10^{-31}$
kg

THE TRIANGLE OF
MASSES

MONSTER NUMBERS
baby steps towards infinity

In an extraordinary work written around 250 BC, Archimedes estimated the number of grains of sand that would fill the then known universe (the solar system) as being 1×10^{63} (in modern notation).

He also realised that for every large number, e.g. the *googol* (10^{100}), there was always a higher order number formed by raising 10 to the power of that number, e.g. the *googolplex* $(10^{googol} = 10^{10^{100}})$. The usual superscript notation becomes clumsy for numbers this big, so let's instead write the googolplex as $10^{\wedge}10^{\wedge}100$. This is a truly monstrous number. We could try and write it as a 1 followed by 10^{100} zeros, but since there are only about 10^{80} atoms in the observable universe, it would be impossible even to write it down.

But what about the next stage up? How big is the *googolplexplex* i.e. $10^{googolplex}$ or $10^{\wedge}10^{\wedge}10^{\wedge}10^{\wedge}2$? Amazingly, numbers of this order have appeared from time to time in respectable academic journals. For example, *Skewes' number* $(\approx 10^{\wedge}10^{\wedge}10^{\wedge}34)$ appeared in 1933 in a paper concerning the distribution of primes and was described by G. H. Hardy as "the largest number which has served any definite purpose in mathematics".

But even the Skewes number is totally insignificant compared to *Graham's number*, **G**, a number so large that it is impossible to write down in our universe, even using the $\wedge$-notation. It is, however, a perfectly computable number and its last 10 digits are ...2464195387.

Recently, mathematicians have come across an even larger number called TREE(3). How close is TREE(3) to infinity? We haven't even started ... TREE(3) is no closer to infinity than the number 2!